I0834244

Technology To Transform Industries and Other Essays

ISBN 978-1-62806-470-4 (print | paperback)
ISBN 978-1-62806-471-1 (ebook)

Library of Congress Control Number 2026904282

Published by
Salt Water Media
29 Broad Street, Suite 104
Berlin, MD 21811
www.saltwatermedia.com

Salt Water
MEDIA

Cover image via istockphoto.com
and used with license and permission

Technology To Transform Industries and Other Essays

Ray Hackert

Contents

Introduction

This is a book of essays, mainly technologies to be developed, according to ideas that I have.

I do not have the time or facilities to develop these technologies, but I encourage readers to pursue them. Some of these technologies will have the capability of revolutionizing industries. Also, they may have the ability to make people's lives better.

For just two examples, if butanol replaces gasoline, travel may be safer and more efficient. And if thorium reactors replace uranium, coal, or natural gas power plants, energy production may be safer, more resilient, and less dirty.

Besides my ideas of technologies to develop, I have added some essays on various other subjects, such as myths I have come across in reading about many things over the years.

Technology To Do

VITAMIN C - AVOID STROKES

Proposal: Investigate the efficacy of vitamin C in preventing strokes

- Summary -

A vitamin C function in the human body is to maintain the health of all the blood vessels. If one takes enough vitamin C each day (I take three grams) all the blood vessels in the body will be elastic and pliable (healthy). This will help avoid strokes.

The alternative, in the absence of vitamin C, is that the blood vessels deteriorate from the inside. This is scurvy, and one dies.

There is no negative to taking vitamin C (except in individuals with diseases that counterindicate it), so amounts up to 2000 mg/day may be taken. Vitamin C is quickly discharged from the body, so it has to be added daily. If injured, through cuts or bruises, blood vessels are all over the body to make repairs where found.

Immune System: The blood vessels have immune cells on their inside walls circulating through them that allow them to help in repairing injuries all over the body. Also, they combat bad active viruses and bacteria.

Collagen: Part of the vitamin C is converted to collagen, a main material for human body structures.

Insufficient vitamin C causes various body problems; these will be discussed in the following sections.

- People's Typical Vitamin C Intake -

Since people normally do not produce vitamin C in their bodies, they get some, but not enough, from food. This means insufficient and wildly varying amounts of vitamin C for people.

Vitamin C's function is to keep the blood vessel system healthy, with elasticity and stretchability. Then it will be the first line of defense throughout the body.

- How Much Vitamin C Is Needed -

Most other mammals besides humans produce vitamin C in their bodies. If a typical adult human wanted to consume the same amount of vitamin C as other mammals make naturally, he or she should take about 1,500 mg per day. Excess vitamin C is excreted from mammals and people through the urine in one or two days. Since people produce no vitamin C, they have to add that much by including vitamin C sources in their diet and/or taking pills. Obviously, if one takes the needed amount by pills, the amount from food is extra; this is no negative and pills cost very little (under 50 cents per day).

- Scurvy -

A lack of vitamin C in people causes scurvy. Scurvy is a terminal disease that affects blood vessels; ultimately, it destroys them from the inside, with bleeding throughout the body.

- Diseases Attributed to Insufficient Vitamin C -

As said before, people cannot usually get enough vitamin C from food such as fruits and vegetables, and the amount they get may vary widely. This leaves them prone to many diseases and symptoms, including poor wound healing, gum swelling and loose teeth, skin spots, minor bleeding, bruising and thickening of outer skin, eyeball bleeding, jaundice, non-normal retina, kidney stones, and non-genetic influence on genes.

- How Vitamin C Can Be Used -

People need to take vitamin C pills to support it at the levels needed. In doing this, they can err on the high side safely, since there is little risk in it affecting body function, and excess is disposed of quickly through the urine.

Mammals other than humans, if injured or under stress, may produce about two or three times their normal vitamin C.

- Insufficient Vitamin C -

Incipient scurvy (my term) happens when the human body has an insufficient and variable amount of vitamin C (ascorbic acid) in the blood system. My hypothesis is that this causes first irritation, and then slight to gross injury of the inside of blood vessels. When the vitamin C level increases somewhat, and the injury to the inside of the blood vessels starts to heal, scar tissue forms at first to clear the injury. If one waits the appropriate length of time, the scar tissue will be replaced by the body (in most cases) with normal tissue (so that the scar is gone). This business of injury, scar tissue, and normal cell replacement is obvious when people get finger cuts, bruises, etc., and a normal healing process. But in the case of ongoing vitamin C insufficiency, repeated and continuous injury of the inside of the blood vessels forms persistent scar tissue over the years. This leaves the blood vessels stiff and inelastic. Could this be a contributor to the arterial plaque that is diagnosed in people?

- Immune System -

I have heard that there are about 60,000 miles of blood vessels in the human body. Immune cells travel throughout blood vessels; they are at every part of the body to prevent attacks by outside sources such as viruses through cuts or bruises, wherever they are in the body. The endothelium lining of the blood vessels helps to regulate the immune cell trafficking. If the endothelium is impaired by scar tissue from vitamin C deficiency, immunity is reduced.

- Body Cell Information -

Cells in the body have vast amounts of information to direct the function of body parts (enzymes, muscles, heart, liver, etc.). Body conditions allow information to be used, or not. Examples include pH or acidity, temperature, and body health status such as sleepiness, tiredness, or weakness. And also, if there is not sufficient vitamin C in the body, that can change the function of an enzyme, for example. Vitamin C acts as a "cofactor," a helper molecule attached to various enzymes, to keep them "reduced," or non-oxidized, so the enzymes can keep doing their jobs.

- Some Chemistry of Vitamin C -

Vitamin C in the body is converted to collagen, a main component of connective tissue. (See section "Diseases Attributed to Insufficient Vitamin C.") It also is an important antioxidant and a regulator of other antioxidants, including vitamin E.

- Historical Vitamin C -

After Columbus, the French, Spanish, and British were warring in the Caribbean to control parts of America. The British had a naval secret. They served a daily ration of sauerkraut or citrus fruit to every man aboard en route to the Caribbean, to prevent loss of people through scurvy. Thus, they could avoid carrying extra seamen expected to be lost to scurvy. Sauerkraut is loaded with vitamin C and can be kept in an open crock jar. But citrus fruit was more reliable and higher in vitamin C, and eventually was adopted as the standard.

- Comment -

It is said that people do not produce vitamin C. I wonder about that. With many billions of people in the world, it would be hard to check them all. I suggest first checking those over 90 years old; I will be the first candidate. Is this far-fetched? No.

With that many people in the world, and each being unique, probably some have developed vitamin C production. I have heard, for instance, that some people are immune to AIDS. Some African people are resistant to malaria. Many mountain populations can live quite well at 10,000 feet above sea level. Why not human production of vitamin C?

- References -

- Meyer, Stephen C., Darwin's Doubts. s.l. : HarperOne, 2014.
- "Vitamin C Fact Sheet for Health Professionals," <https://ods.od.nih.gov/factsheets/VitaminC-HealthProfessional/>, 52 pages

- 3. Cell Press. "How Humans Make Up For An 'Inborn' Vitamin C Deficiency." ScienceDaily. ScienceDaily, 21 March 2008. <www.sciencedaily.com/releases/2008/03/080320120726.htm>.
- Maxfield L, Crane JS. "Vitamin C Deficiency". [Updated 2022 May 8]. In: StatPearls [Internet]. Treasure Island (FL): StatPearls Publishing; 2022 Jan-. Available from: https://www.ncbi.nlm.nih.gov/books/NBK493187/

ULTRASONIC ENERGY USES

Proposal: Investigate new applications for ultrasonics, such as improving iron ore yield, producing ammonia, sterilizing milk, and more.

Ultrasonics is a generally available source of high energy treatment.

Ultrasonic energy in water is observable by its cavitation. The word cavitation means that sound, when at very high frequencies, vaporizes a few molecules of water (millionths of a gram) way beyond its boiling point to about 1300°C. Then, being surrounded by room temperature water, it collapses back to room temperature giving off a light that indicates the high temperature and high energy involved. Therefore, ultrasonic energy can be used for both high temperature treatment as well as high energy treatment. This should be (and is) useful to promote all kinds of chemical and physical reactions.

- My Experience with Ultrasonics -

Ultrasonic treatment of oils in water gives a microemulsion (transparent even at 30% oils). Microemulsions like these are added to molten 6,6 nylon as it spins out and is frozen into a bundle of filaments, a threadline. Each filament in the threadline is completely covered with microparticles of lubricating oil in water. Good lubrication is necessary since the yarn is moving over pins, guides, and rolls at maybe 1300 meters per minute, and is being drawn out to typically 5 times its spun length.

TiO_2 (Titanium Dioxide) Dispersion: Ultrasonic processing is key to producing quality nylon because of its ability to disperse each one micrometer TiO_2 particle in water. Ultimately each particle is dispersed in molten nylon and spun through spinnerets with holes through them that are very small, about the size of a human hair. All 6,6 nylon manufactured is "glassy" nylon and would be transparent without the TiO_2 particles, which make them look white. Ultrasonic dispersion of TiO_2 in water has to be controlled to a specific energy level. If too much, the coating on the TiO_2 particles is ripped off and a trace of iron atoms in the TiO_2 particles is exposed to the nylon. The iron atom contamination catalyzes nylon gel particle formation, and it results in shutting down the nylon factory's process – with plugged spinnerets at worst, or at least, poor quality, nonsalable yarn. We ran an experiment and got these results.

Spinneret Cleaning: Ultrasonics are used to clean nylon spinnerets. Spinnerets are blocks of steel with micro-holes in them, used to extrude molten 6,6 nylon at 280°C and typically between 10,000 and 35,000 psi. After they are used, the spinnerets go through a burning process to remove nylon residue. Then they are placed in a tray with cleaning solution in it, and ultrasonically cleaned to remove any particles left in the micro-holes of the spinnerets.

Other Uses of Ultrasonics from My Lab Experience as an Analytical Chemist

- To analyze a sample with microbes in it, one treats it with an ultrasonic probe to destroy the microbe's cells so the cell contents are available to also be analyzed with the bulk content.
- In water and waste analysis, an ultrasonic probe is also used to sterilize these samples as needed. For example, in studying ways to biodegrade a chemical with a new microbe, one has to know that the only microorganism in a sample is the microbe seed that we are testing.
- The ultrasonic probe is used in the lab to generate data to commercialize ultrasonics in factory conditions (scaled from very small lab experiments to 20,000-gallon tanks, for example). The probe is available from lab equipment companies.

- Potential Applications -

Ultrasonics has great potential to treat various materials with high temperature and high energy to convert them to useful products cheaply.

Iron Ore: Many experiments show some success using high energy applications, such as ultrasonics, to convert hematite (industrial waste) to magnetite (the component used commercially) in iron ore. If successful, mining iron ore would be far more productive, since many ores do not have much magnetite in them to begin with. Most iron ore companies now magnetically separate out the magnetite compounds and ship them to steel mills. If a cheap way is developed to convert hematite iron compounds to magnetite, even lousy ores could become profitably mined. Processing using ultrasonics could be that inexpensive method of increasing usable product from the ore. (See references.)

This is not farfetched. Titanium dioxide mining, to be commercially attractive, had to find and use rich ores containing it. A technology was developed to convert the TiO2 in ores to titanium tetrachloride, a gas, and then separate the titanium.

Nitrogen Fixation: A simple cheap method is needed to replace the Haber-Bosch process for producing ammonia for fertilizers and other products. Some small success has been achieved by using an electric arc. (Nature probably produces ammonia in air by lightning strikes.) Ultrasonics may provide an even better process for ammonia production.

Nylon Polymerization at Room Conditions: Ultrasonic treatment of 6,6 nylon salt (48%) in water is expected to give nylon polymer at room conditions.

Nylon water solution is the starting point to commercially produce nylon in autoclaves, using high temperatures and pressures. In the ultrasonic treatment, at the point of cavitation, very high temperature and pressure occur in a micro-space and micro-time, and leave the bulk of the liquid mainly unchanged. This probably would produce low-molecular-weight nylon very cheaply, that then could be further processed for special end uses.

Medical Applications: 2022 articles (see below) tell about using ultrasonics to glue things together in the human body.

- References -

Iron Ore

- Pasechnik, L. A., Skachkov, V. M., Bogdanova, E. A., Chufarov, A. Y., Kellerman, D. G., Medyankina, I. S., & Yatsenko, S. P. (2020). A promising process for transformation of hematite to magnetite with simultaneous dissolution of alumina from red mud in alkaline medium. Hydrometallurgy, 196, 105438.
- Sahu, S. N., Baskey, P. K., Barma, S. D., Sahoo, S., Meikap, B. C., & Biswal, S. K. (2020). Pelletization of synthesized magnetite concentrate obtained by magnetization roasting of Indian low-grade BHQ iron ore. Powder Technology, 374, 190-200.
- Zhao, B., Gao, P., Tang, Z., & Zhang, W. (2021). The efficient improvement of original magnetite in iron ore reduction reaction in magnetization roasting process and mechanism analysis by in situ and continuous image capture. Minerals, 11(6), 645.

- Mitov, I., Stoilova, A., Yordanov, B., & Krastev, D. (2021). Technological research on converting iron ore tailings into a marketable product. Journal of the Southern African Institute of Mining and Metallurgy, 121(5), 181-186.
- Sun, Y., Zhang, X., Han, Y., & Li, Y. (2020). A new approach for recovering iron from iron ore tailings using suspension magnetization roasting: A pilot-scale study. Powder technology, 361, 571-580.
- Gaviria, J. P., Bohé, A., Pasquevich, A., & Pasquevich, D. M. (2007). Hematite to magnetite reduction monitored by Mössbauer spectroscopy and X-ray diffraction. Physica B: Condensed Matter, 389(1), 198-201.
- Ponomar, V. P., Brik, O. B., Cherevko, Y. I., & Ovsienko, V. V. (2019). Kinetics of hematite to magnetite transformation by gaseous reduction at low concentration of carbon monoxide. Chemical Engineering Research and Design, 148, 393-402.
- Matthews, A. (1976). Magnetite formation by the reduction of hematite with iron under hydrothermal conditions. American Mineralogist, 6, 927-932.
- Cavanaugh, P. (1959). Method of converting hematite to magnetite (U.S. Patent No. US2870003A). U.S. Patent and Trademark Office.
- SHANGHAI MAIFENG MICROWAVE EQUIPMENT CO Ltd (2011 – patent pending).

Method for converting natural non-magnetic iron ore into magnetite (China Patent No. CN102168170A).

- Ponomar, V. P., Dudchenko, N. O., & Brik, A. B. (2018). Synthesis of magnetite powder from the mixture consisting of siderite and hematite iron ores. Minerals Engineering, 122, 277-284.
- Xiong, D., Lu, L., & Holmes, R. J. (2022). Physical separation of iron ore: magnetic separation. In Iron Ore (pp. 309-332). Woodhead Publishing.
- Yu, J., Han, Y., Li, Y., Gao, P., & Li, W. (2017). Mechanism and kinetics of the reduction of hematite to magnetite with CO–CO2 in a micro-fluidized bed. Minerals, 7(11), 209.

Medical Applications

- Es Sayed, J., & Kamperman, M. (2022). Ultrasounding out a technique that sticks. Science, 377(6607), 707-708.
- Ma, Z., Bourquard, C., Gao, Q., Jiang, S., De Iure-Grimmel, T., Huo, R., ... & Li, J. (2022). Controlled tough bioadhesion mediated by ultrasound. Science, 377(6607), 751-755.

SCIENTIFIC INVESTMENT (BUSINESS PLAN)

Proposal: Encourage individual investors to use methods that are scientifically sound.

When investing, one should have a plan to invest wisely; this plan given below will aid one to be efficient and effective.

- Buy low and sell high, or keep the investment.
- Buy into stocks you know, or industries you know.
- For long term, I have more faith in stocks than bonds. With stock, when you buy it, you have it as long as you want it. With bonds, the investment is generally for a finite time.
- Capital gains is a tax on a stock investment; it is the difference between the price paid for it and the selling price. This capital gains can be avoided by donating the stock at its current market value and buying a replacement share of stock. An example is when one goes to sell a share of stock with price paid of $50 and price to sell at $70 with a capital gain of $20. To avoid the $20 tax, one donates it and purchases a replacement share.
- In a deep depression, the grain industry (food) should be more reliable than other investments.
- For grain companies to make more profit, they are edging into the chemical business – for example:

converting their corn to corn syrup that is being used in vast amounts (2020) in soft drinks. (In this form, they are taking advantage of the significant consumption of corn by humans.)

- In a depression, or in a stock bankruptcy, a manufacturing company has plants and inventory as a continuing cost because they are sitting unused; grain companies, on the other hand, maintain their inventory and assets which have continuing value – the inventory is food, and that is needed by people even in a depression – people MUST EAT.
- In 2020, grain company stocks were valued by the market at about price-to-book value of one (1). So if a grain company goes broke, the company assets are approximately equal to the value of the company. My opinion: if a grain company goes into bankruptcy, others will buy it; grain companies operate on a small margin of profit in vast volumes. The grain business is very competitive. That keeps its margin low, so to make more profit, they need more volume.
- Be ready for recessions and depressions to buy low. Buy low whenever it happens (such as 9/11 or COVID-19).
- If a significant amount of a stock is held by institutions such as insurance companies or universities, with their vast resources to evaluate a stock, they generally will invest in good, reliable stocks. Learn

about who holds the stock you're interested in before investing.

- Let the market teach you what is a good stock.
- In the 1929 Depression, second and third derivative investments were lost. An example of a second derivative investment is one displaced from the original owner by two steps. For example, the stock is in a stock fund of company A (first derivative) and company A is in another fund of first derivatives stocks, company B (the second derivative). During the Depression, company A and company B failed, but the company that held the original stock was less likely to fail. Therefore, buy stock with low or ideally derivatives.
- Listen to what company officials are buying or selling of their own stock. They have inside information of their companies before it is known to the public, allowing them to make wise investment decisions, but they must report this to the government. Judge for yourself.
- There are many company characteristics, such as share-price-to-value ratio, that are and have to be reported frequently, so be a wise investor and see how they compare to other companies, and how the company is doing. Other company characteristics to look at are debt and debt-vs-value.

- There is always the possibility of improving one's investment portfolio. Always be on the lookout for a better investment.
- While money in a bank account is very liquid, nowadays it makes very little interest. Hold "cash" in the form of investments that pay interest or dividends but are readily sellable, so you can buy into good opportunities quickly.
- Buy land if it is a good investment. Stock companies come and go, but land is "permanent." (I have none, since it is too pricey, now in 2024.)
- Exercise patience. For example, figure out what investment is likely to drop in price, and then WAIT until it drops before buying it. But have lots of information – there's no purpose in waiting if you don't know what to expect.
- Everything is negotiable. For example, you can ask your broker to keep an eye on the price of a stock and buy it only if it reaches the price you are seeking.
- Keep up with what might be happening in the market by following the information in newspapers, TV, etc.

- Comments -

In the 1970s, interest rates were at about 17%, and they were coming down. A savings and loan I worked with avoided bankruptcy but were slowly working out their cash-flow problem. They were asking people to invest $25,000 or more, and I offered them $40,000 and got an agreement to pay me 12% for five years. As I remember, the declining interest rate was already down to about 9-10%. The interest I got through this deal was used to aid in my youngest daughter's college education.

Another example is that during the COVID-19 low market, I bought Anderson Grain Co. stock for an average price of $11 a share. In two years, it was back up to its pre-pandemic price of $35-45 a share. 3000 shares gained over $20 a share.

NITROGEN FIXATION VIA FERMENTATION

Proposal: Investigate ammonia production through fermentation.

There is a lot of work going on (as of October 2022) to achieve cheap nitrogen fixation. For example, they use microbes found in the roots of legumes; these fix nitrogen from the air to make ammonia fertilizer in various ways. The microbes are added dry to corn kernels, and they are included in micro-packets to aid fertilization of various crops, with some success.

Obviously, these nitrogen-fixing bacteria are very rugged and lend themselves to various forms of handling while continuing to be effective.

One should be able to modify these to make fermentation microbes that would then be used to make ammonia. This would have tremendous agricultural and financial advantages. A major cost of commercial farming is fertilizer. It is costly because it is made by the Haber-Bosch process in big factories at high temperatures and pressures.

Fermentation is already known in agricultural areas – for example in ethanol production. Ethanol production is subsidized by the government. In comparison, fermentation to produce cheap ammonia as fertilizer would be cheap and self-sustaining, and might be done in small farm- or community-sized fermentation plants near where fertilizer is needed and used, on the farm. This would revolutionize agriculture.

In the future, I would expect fermentation to also make urea. Today, this is the slow-release fertilizer of choice. Urea is a stable solid, as opposed to ammonia, which is a gas that normally is used in a water solution.

Grain companies already are proceeding past fermentation-generated ethanol to more profitable and more complex chemistries by fermentation (for example, Cargill in Central Iowa). Others should be doing the same.

THORIUM (Th)

Proposal: Commercialize small thorium reactors as a safer, cleaner, more secure energy source

Thorium (Th) is available as a better alternative to uranium, which is used now for generating electricity all over the USA.

A thorium reactor was first developed in the 1930s as an early possibility to develop atomic bombs and then hydrogen bombs, in competition with Nazi Germany's work. Thorium was dropped because one cannot make hydrogen bombs this way.

Today, atomic energy for electricity uses uranium, a spinoff of the extensive experience with it for bomb development. Thorium is better suited to supply energy for civilian and industrial needs, since it is considerably safer and much simpler to use for energy supply.

Thorium process waste is composed of less dangerous and shorter-lived components than from uranium power plants, including significantly less plutonium; plutonium has a half-life of about 24,000 years (impossible for humans to deal with).

Another favorable characteristic of the thorium reactor – it was self-regulating: the more energy was removed, the more energy was made. This was not the case with uranium. It needs an extensive, complicated control; for example, operators would need to move the uranium-rich bars in and out to adjust energy output, using an electromechanical system. And one of these mechanical systems that can fail.

There is development going on in government labs to achieve small thorium reactors to power space missions (https://science.nasa.gov/planetary-science/programs/radioisotope-power-systems/about-plutonium-238/). One can expect that to move into cars, trucks, boats, etc.

I think that thorium should be adopted as the main source of USA energy. As a chemist, I think of petroleum and coal as black gold, to be used for many things useful to humans, such as clothing (fibers) and building materials, not to be just burnt for its energy content to move a car or heat a house.

Thorium is plentiful all over the earth. I have heard people claim there is enough in most people's yards to supply their energy needs for their lifetimes. In fact, it is said that thorium is prevalent throughout the earth; its radiation is what keeps the earth's center hot enough. Otherwise, the earth would be a frozen ice planet.

Currently there is extensive government regulation of atomic energy (that is, of the uranium process) because of its many hazards as shown over the years. The uranium processes, even with their comprehensive and complete regulations, have given us disasters like Chernobyl, Three Mile Island, and Fukushima. In comparison, a thorium process is and can be fail-safe. For example, if overheating/overreacting occurs, a temperature-sensitive plug melts and the reactor contents flow out into a container and freeze.

A particular advantage of thorium is that it operates dissolved in a molten mixed salt at about 800°C. This hot system can be used directly for high temperature industrial needs like producing cement, or in steel mills, where now electric heating is used. The system could bypass coal or gas combustion to make electricity, to send it into the grid to use it for high-temperature processes, but instead, it would be more efficient to use the heat right where it's produced.

Do we in the USA know how to do thorium processes? Yes! A number of years ago, I saw on TV where the US government was encouraging Iran to stop their bomb development, since they claimed they needed uranium for medical purposes. The USA told them we would give them a process for creating nuclear energy. We would give them the core materials and process (know-how) and would set them up with our people helping them. Iran rejected this; obviously they wanted a bomb.

For further reading, see below.

- Reference -

- Hargraves, Robert (2012). Thorium: Energy, cheaper than coal.

IMPROVED SUNSCREEN

Proposal: Investigate manganese hypophosphite as an improved sunscreen.

Manganese hypophosphite has a good chance of beating commercial sunscreen, in both cost and function.

This compound has been used as a 6,6 nylon stabilizer for many years. 6,6 nylon without this stabilizer disintegrates in sunlight. When carpet 6,6 nylon was first being developed, the stabilizer was left out (by mistake) and the test carpets in motels in southern Florida fell into shreds when they were exposed to sunlight.

I suggest that, since nylon is made of polyamides and so is human skin, manganese hypophosphite's ability to prevent destruction of nylon-based carpet fibers may also help prevent sunlight's destructive effect on skin.

My proposal is to add a trace of manganese hypophosphite to skin creams and test its function as a sunscreen. These tests should first establish an optimum functional concentration of use. Further testing will be needed for human safety and efficacy.

AUTOMOBILE FUEL

Proposal: Commercialize butanol as a safer, cheaper vehicle fuel

- Gasoline to Butanol -

Historically, gasoline was introduced in the early 1900s by the infant petroleum industry as a better petroleum fraction compared to kerosene, to expand business. It soon was introduced for use in the internal combustion engines, for passenger cars and trucks. The following description of gasoline problems and butanol advantages follows.

- Background -

Please note the articles listed in "References" about butanol technology. They will lead to much more information in the open literature, including patents that most likely could be used for a competitive, reasonable price.

- Advantages of Butanol Over Gasoline -

- Safety – Butanol is a low-vapor liquid, which gives it minimum concerns in handling (like fueling your car, lawnmower, or boat). Gasoline is a very hot-burning explosive. (Remember Molotov cocktails in World War II and 500-pound fire bombs.) The NASCAR industry should look at butanol to replace gasoline and ethanol – gasoline has caused fires in crashes, killing star drivers, particularly young ones.

- Isobutanol, as it forms in the water medium of fermentation, is not "hygroscopic" (water-absorbing), and separates easily – no costly separation process is needed. Ethanol, on the other hand, is prepared by fermentation so you have water and ethanol. Then you have to distill off the ethanol from the water and that's where the cost is and it can be very expensive.
- Butanol, when partially replacing gasoline in a gasoline-engine car, has been reported to get almost the same miles per gallon (MPG). This has happened in many tests in the last several decades, even though butanol has 75% of the burnable carbon and hydrogen compounds (energy) as gasoline. See "The Problems with Gasoline Engines."
- Butanol, in an engine built for it, has (at least) the potential for a 20-30% MPG efficiency improvement. I also predict that will come on soon.
- Butanol will not need an antiknock agent in fuel for engines. (Ethanol is antiknock agent for gasoline.)
- Isobutanol is moderately soluble in gasoline. Therefore, transition to butanol from gasoline can be freely done. Gasoline transfer systems can be used – with no new infrastructure, not like ethanol use where it required special handling because it absorbs water (moisture) from air.

- One of the most fruitful places where butanol should replace gasoline is in boating or other vehicles where fires could be especially deadly. The technology is available, and many major companies and governments are into it already.
- Butanol is, and will be, cheaper than gasoline. While gasoline needs petroleum to produce it in large but few refineries, fermentation production of butanol all over the USA (and the world) will minimize its cost to deliver to the consumer. Many sources of butanol vs. few sources of gasoline guarantee minimal upset of butanol use in a crisis (natural or due to man).
- Butanol (like ethanol) will continue the trend of distribution of the industry throughout the whole country. (Local industry for local populations – wealth does not go off to major rich city organizations).

- Butanol Concerns -

In the early use of yeast-type enzymes/catalysts to ferment butanol, butanol poisoned the microbe population. As one way to deal with this, butanol producers have adapted a time-tested chemical engineering technique – as the insoluble butanol is formed, it is removed, leaving the microbe population safe and ready to produce more butanol. More recently, gene-modification scientists have found a gene that causes the poisoning in yeasts, and removed it.

- The Problems with Gasoline Engines -

In my opinion, gasoline engines are about at their limit in correcting for the defects in using a very hot-burning explosive like gasoline as an automobile fuel. In order to get more energy out of gasoline engines, they have gone to higher compression ratios, making the problem worse. In order to prevent engines from getting too hot (and being destroyed in the process), gasoline is used to cool the engine. This is worsened further then by having engines running rich (too much gasoline, not enough air) and inefficient.

Also, gasoline engines knock if one does not use an antiknock agent such as tetraethyl lead (as in "leaded gas" in the past) or an alcohol (such as ethanol).

Running hot, gasoline engines are famous for making oxides of nitrogen and also unburned carbon compounds in their exhaust to pollute air. Corrections of these problems use a catalyst in the exhaust system. All of this environmental equipment robs energy for moving cars. Because butanol burns cooler, it should make far less oxides of nitrogen, requiring little or no environmental equipment.

In conclusion, the world should use butanol as automobile fuel, ultimately in butanol-optimized engines.

And although butanol is a promising new automobile fuel, I recommend that a research effort be done to find the ideal fuel, and then take the steps needed to switch to it.

- References -

- "New MOF can efficiently separate biobutanol from fermentation broth," 29 April 2020 https://www.greencarcongress.com/2020/04/20200429-mof.html
- "Gevo and Leaf Resources sign joint development agreement for potential use of cellulosic-derived sugars to convert to hydrocarbon molecules," 12 September 2019 https://www.greencarcongress.com/2019/09/20190912-gevo.html
- "Gevo to deploy Shockwave process to lower the carbon intensity of its ethanol, isobutanol," 15 August 2018 https://www.greencarcongress.com/2018/08/20180815-gevo.html
- "NUS team discovers bacterium that produces only biobutanol directly from cellulose," 06 April 2018 https://www.greencarcongress.com/2018/04/20180406-nus.html
- "Princeton team uses light to boost production of isobutanol 5x from engineered yeast; optogenetics," 23 March 2018 https://www.greencarcongress.com/2018/03/20180323-princeton.html
- "ORNL/TM-2013/243 Compatibility Study for Plastic, Elastomeric, and Metallic Fueling Infrastructure Materials Exposed to Aggressive Formulations of Isobutanol-Blended Gasoline," August 2013 https://info.ornl.gov/sites/publications/files/Pub44488.pdf

- "The Top 10 advances in renewable butanol: what's speeding up, where are the slow-downs?" August 2, 2018, Jim Lane http://www.biofuelsdigest.com/bdigest/2018/08/02/the-top-10-advances-in-renewable-butanol-whats-speeding-up-where-are-the-slow-downs/
- "Jet Fuel, Isooctane Gasoline, Ethanol and Isobutanol: The Digest's 2020 Multi-Slide Guide to Gevo's Low Carbon Drop-In Transportation Fuels," May 7, 2020, Jim Lane http://www.biofuelsdigest.com/bdigest/2020/05/07/jet-fuel-isooctane-gasoline-ethanol-and-isobutanol-the-digests-2020-multi-slide-guide-to-gevos-low-carbon-drop-in-transportation-fuels/
- "Princeton researchers find fundamental mechanism of isobutanol's toxic effect," April 28, 2020, Meghan Sapp http://www.biofuelsdigest.com/bdigest/2 020/04/28/princeton-researchers-find-fundamental-mechanism-of-isobutanols-toxic-effect/
- "International researchers develop metal organic framework for biobutanol production," April 27, 2020, Meghan Sapp http://www.biofuelsdigest.com/bdigest/2020/04/27/international-researchers-develop-metal-organic-framework-for-biobutanol-production/
- "Seaweed-based biobutanol proven in road test," January 1, 2020, Jim Lane http://www.

biofuelsdigest.com/bdigest/2020/01/01/seaweed-based-biobutanol-proven-in-road-test/

- "Engineered Strains for Fermentation: The Digest's 2019 Multi-Slide Guide to n-Butanol from Biomass and CO2," December 3, 2019, Jim Lane http://www.biofuelsdigest.com/bdigest/2019/12/03/engineered-strains-for-fermentation-the-digests-2019-multi-slide-guide-to-n-butanol-from-biomass-and-co2/
- "SilicoLife's biobased n-butanol IP advances," October 27, 2019, Jim Lane http://www.biofuelsdigest.com/bdigest/2019/10/27/silicolifes-biobased-n-butanol-ip-advances/
- "Seattle successfully complete phase 1 of pilot trial using Gevo's isobutanol," August 8, 2019, Meghan Sapp http://www.biofuelsdigest.com/bdigest/2019/08/08/seattle-successfully-complete-phase-1-of-pilot-trial-using-gevos-isobutanol/
- "Uppsala University produce butanol from cyanobacteria," July 29, 2019, Meghan Sapp http://www.biofuelsdigest.com/bdigest/2019/07/29/uppsala-university-produce-butanol-from-cyanobacteria/
- "Green Biologics closes Minnesota butanol and acetone plant," July 14, 2019, Helena Tavares Kennedy http://www.biofuelsdigest.com/bdigest/2019/07/14/green-biologics-closes-minnesota-butanol-and-acetone-plant/

- "Evonik and Siemens launch phase 2 of Rheticus: butanol, hexanol from CO2 and water using renewable electricity and bacteria," 20 October 2019 https://www.greencarcongress.com/2019/10/20191020-rheticus.html
- "U Michigan team develops air-stable catalyst for upgrading ethanol to 1-butanol," 04 January 2016 https://www.greencarcongress.com/2016/01/20160104-umich.html
- Five bio-based companies to watch, Kapil Shyam Lokare, PhD https://www.fuelsandlubes.com/fli-article/five-bio-based-companies-to-watch/
- Bio-based Butanol Industry Size 2020 by Manufactures Types, Applications, Regions and Forecast to 2024 https://www.wrcbtv.com/story/42082421/bio-based-butanol-industry-size-2020-by-manufactures-types-applications-regions-and-forecast-to-2024
- Kalita, M., Muralidharan, M., Subramanian, M., Sithananthan, M., Yadav, A., Kagdiyal, V., ... & Suresh, R. (2016). Experimental studies on n-butanol/gasoline fuel blends in passenger car for performance and emission (No. 2016-01-2264). SAE Technical Paper.
- Huang, H., Liu, H., & Gan, Y. R. (2010). Genetic modification of critical enzymes and involved genes in butanol biosynthesis from biomass. Biotechnology advances, 28(5), 651-657.

Political To Do

INCOME TAXES

Proposal: Replace the US income tax with Fair Tax on purchases.

It seems strange to me that there is a corporate income tax. Corporations (and businesses in general) pay their taxes and write it off as a business expense. These taxes (and fees) show up in the business costs, so people buying their products are paying the taxes passed on to them. I have never estimated how much of that tax I pay when I buy a loaf of bread, an apple, or a chair. Just thinking about it, it seems to me that it would be up to 50% of the price of anything bought. If one includes that with direct taxes like personal income tax, we are very heavily taxed.

I have heard of a new tax in Europe that is added to all the other taxes, the "fair" tax and is just another sales tax. My opinion is that the income tax was an amendment to the US constitution and should be repealed and done away with, and only have the "fair" tax. This would do away with the IRS department as we know it. Instead of having to file an income tax return, people would pay their tax when they bought something.

For almost anything I do now, I have to think about how it affects me with the IRS. Life would be much simpler if there was no income tax, only a "fair" sales tax.

Another aspect of the IRS is that it assumes you are guilty until you can prove your innocence in a court. The

IRS says you are guilty, then starts issuing fines! A $5,000 mistake quickly becomes $20,000 or $50,000, with no way to control the IRS' actions. Try to litigate that in a court, usually after the IRS has already taken your money or your business (without a court process)! The IRS has acquired dictator powers not given in the constitution.

I have seen a family business maintaining a suspense account just to cover IRS interpretation of the income tax rules vs. their own tax lawyers' interpretation; it's not worth taking a $3,000 IRS judgment to go to court and pay maybe $5,000 or $6,000. I call that situation "fuzzy" regulations.

With the fair tax added (and the income tax removed!), the IRS would not come after individual citizens about whether they had paid the correct tax. This (hopefully) simplified burden would fall on those who sell goods and services.

SCHOOL CHOICE

Proposal: Allow families to choose alternatives to their local public school for their children, with accompanying tax credit.

Public schools are run by the government and are monopolies. They are controlled by the teachers' unions (an arm of the Democrat/Socialist party). At worst, they teach Democrat/socialist ideas and principles; this time should be used teaching and educating the children for dealing with life. To combat the monopoly, tax money for education should go with individual students, and their families should have a choice about which school they want them to go to.

In my home, Wicomico County, we already have a number of private schools, but most students' families cannot afford them. This is why school tax money should go with the student to pay for his school.

In Virginia, their governor Glenn Youngkin was elected, and he was advocating school choice. School choice is politically important to voters.

If school choice is adopted, this would hopefully wipe out the current Democrat/Socialist teachers' unions and their gross effect on public schools.

If school choice is adopted, schools can hire teachers with proven capabilities, not just union teachers. In addition, capable teachers should be paid as well as engineers. After all, they have a great responsibility to teach our kids.

We don't want stupidity prevalent in our adult population. In comparison, just as we expect knowledge and quality work from our engineers and chemists, we should expect these from our teachers as well.

US CURRENCY

Proposal: Reduce US debt and return US currency to a stable standard.

In 2023, for the very first time, the Fitch US credit rating was downgraded from AAA (triple A) to AA+. This probably was caused by excessive USA debt. We have become a welfare state, promising and paying more and more for welfare programs, using borrowed money. The amount of money that can be borrowed is limited by the lender, just as it is with individual borrowing. This amount can become excessive; but the lender will limit borrowing when they see that the borrower cannot possibly repay the loan. If the USA borrows too much for these welfare programs, the lender could cut us off, putting an end to these programs.

In 1913, the USA passed a law to introduce the Federal Reserve. Rather than being able to maintain a stable value of money as with a gold standard, its goal was 4% inflation. The idea was to have a system where you paid debts for borrowed money with always cheaper and cheaper money. This encouraged debt. The question is: How much is realistic? Already, a good part of our tax income is needed to pay off interest on the country's debt. France borrowed too much and now is sanctioned by the EU against borrowing, and downgraded. If we keep going down that same road, we eventually will not be able to borrow any more. Our welfare state will be very restricted on what it can borrow.

Meanwhile, everything keeps going up, so the states will have to cut back. People will have to reduce their living standard. Will they do it willingly? Probably not.

SUBSIDIES

Proposal: Stop subsidizing desired technology development; let capitalism drive innovation.

Politicians, wanting certain technologies to be developed commercially, use their political power to pass subsidies in Congress that incentivize people to use these technologies. This should be opposed, since my taxes should not be used for politicians' own purposes. It is not up to politicians to choose an economic outcome. Let capitalism and the economy make that decision.

One illogical subsidy is for windmills used in generation of electricity. Windmills are controlled to a maximum speed so that the blades do not get destroyed by them going too fast. A windmill's efficiency is limited by the strength of the material that is used for the propeller blades. Windmill power is not commercially viable in the sense that windmills' electricity adds complication to the accounting process for the grid. Its production is variable and sporadic based on wind conditions, and so is not a simple, steady addition of electricity to the grid. While wind power is heavily subsidized, so are coal and natural gas power. I recommend to cut the subsidies, and "let the best approach win," driven by the marketplace and capitalism.

Another subsidized industry is electric cars and gasoline/electric cars. They are very inefficient: They use grid electricity, which is already very inefficient. It loses about one third through transmission losses, since it is alternating current. (Direct current (DC) on the grid would have had far less loss in transmission, but that's an argument for another day.) This problem has already been solved for railroads. They use diesel engines on locomotives to generate DC electricity, which then goes to a DC motor to drive the engine. There is no reason I can see that railroad technology cannot be adapted efficiently to cars and trucks. We only need to size it down from railroads to cars.

But gasoline is not very practical for running cars. Much of cars' efficiency is lost in using gasoline, not to mention that it is a fire hazard and an explosive material.

Some people have been working on butanol for a transportation fuel. Testing has shown that it gives about the same mileage (mpg) as gasoline in a gasoline car, but is very safe in comparison to gasoline. Butanol has about 70% of the carbon as in gasoline, but if the engine does not need the environmental gear that gasoline requires, an engine designed for it should give at least 40 or 50 mpg. But to be scientific, one should identify an optimum transportation fuel and adopt it. This should be compared to diesel, already in use for transportation fuel.

Again, do not subsidize this progress. Let capitalism and the market do it. This will take time, but the goal is the best transportation fuel.

WEIRD LAWS

Proposal: Write laws and regulations so that they only kick in if and when their underlying assumptions are true, and sunset them if the assumptions prove false.

Generally, the problem is laws that are passed that are mainly wishes. Several examples are given below.

- Ethanol -

The law was to mix ethanol into gasoline. Ethanol is needed as an anti-knock agent for gasoline – to replace other complex and costly chemicals such as tetraethyl lead (known as leaded gasoline). The law's goal was to get ethanol in gasoline to 85% (E85). However, we came to find out, after being in this program, that above about 10% ethanol in gasoline was possible, but it corrodes gasoline engines. (The exception, probably, is Ford, which makes cars for Brazil to run on ethanol alone.) Of course, car manufacturers and car and truck owners are against the destruction of their vehicles.

Another negative for ethanol in gasoline is that except for very low concentrations of ethanol, ethanol/gasoline has to be handled separately from gasoline: Ethanol picks up water from air and one does not want water in gasoline.

An alternative to ethanol as an anti-knock agent in gasoline is to use butanol. Butanol is like gasoline in that it does not dissolve water from the air. Butanol is completely soluble in gasoline, so any amount can be added to gasoline without special handling like ethanol in gasoline. Butanol has the advantage over gasoline in that it has a very low vapor pressure, so it is very safe during handling. In fueling with butanol, it will not ignite, so sparks or lit cigarettes will not cause butanol to explode like gasoline will. Car fires mostly would be eliminated using butanol.

In the past, many have evaluated butanol in cars built for gasoline. Ultimately, if butanol is adopted as a transportation fuel, engines should be designed for this fuel. Negative problems for gasoline which compromise gasoline's efficiency can be avoided, and one should get at least 40, maybe up to 60 miles per gallon.

- Solar and Wind -

Politicians have chosen to encourage these forms of electricity production, by giving a government subsidy for using them, and forcing electric power companies to accept this more costly electricity. Customers pay for this more costly electricity (solar and wind), both for its cost, and for its handling by regular, traditional electric companies which were not designed with wind and solar in mind. Both solar and wind technologies are a long way from being financially competitive. Their technologies are so poor compared to coal or petroleum or gas electricity production that they never will be cost effective, at least for the next 30 years or more.

An example is that as far as electricity generation by solar is concerned, it's been calculated that today, solar technology would require all of Nevada to be covered by solar cells, to supply only Los Angeles with electricity – that is how little solar can do today. And wind is no better.

- Electric Vehicles (EVs) -

There are a lot of laws and rules passed where the government subsidizes electric vehicles. I do not believe that the US government should subsidize development of new technology, against the principles of capitalism.

Where does the electricity come from for EVs? From the electric grid, of course. The grid electricity is one of the worst polluters, if you think CO2 (carbon dioxide) is pollution. Think of it. There is a certain energy loss to burn coal, gas, or petroleum to make electricity. Then to transmit electricity on the grid has a certain energy loss. Then lastly, use of the electricity in the vehicle also has energy loss.

Let's make an example. Let's assume each step loses 50% of the energy it receives – so lose 50% to make it, 50% of what's left to transmit it, and 50% of what's left to use it. This gives 50% of 50% of 50%, or 12.5% of the energy you started with is left to move the car.

If a person just uses gasoline in one car, for the same logical reasoning, that would give 50% efficiency for gasoline cars vs. 12.5% for electric cars.

Another form of reasoning: Pure electric cars using the

grid example above would be less efficient than gasoline/ electric (hybrid or rechargeable hybrid) cars, particularly if they ran mostly on gasoline.

Besides, electric cars of any sort are quite costly, so with all the comments on efficiency above, to drive a more costly car adds to one's cost-per-mile to drive. Of course, that is one reason people drive old cars: They have been depreciated to zero cost, so driving costs include only fuel and repairs.

- Thorium (Th) -

There is a lot of law now on the books in the USA about atomic energy – all kinds of rules and regulations. But when we come up with a new technology like thorium-based atomic energy, the lawyers will try to apply the old laws in ways that don't make sense to the new process, putting obstacles in the way of implementing this innovation.

Thorium for all energy production in the USA would be expected to raise everyone's standard of living. That would be good for the country, since for years, we have been having more and more poor people.

Other Essays

<u>CARBON DIOXIDE</u> (CO_2)

- Climate Change -

Climate change is claimed to be caused by carbon dioxide (CO_2) discharge into air by industries' burning of fossil fuel; this causes global warming. My theory is that the world has been warming since the Little Ice Age of about 1750. As the seas warm up, soluble gases such as CO_2 in them become less soluble and discharge into the air.

The seas cover 70% of the earth. CO_2 discharged into the air from the seas dominates the CO_2 in air; industrial CO_2 into the air is a small part of the CO_2 in air.

Climate changers have been debunked in that, for decades, they have predicted dire results like sea levels rising a foot or more in a decade or two, and it has not happened. (Ocean rise is estimated at about 1.4 inches per decade.)

Climate changers have achieved financial support to encourage their claims. They have set up markets in Europe and the USA where big "polluters" pay for their large CO_2 discharge, and minor "polluters" get paid off, with market fees going to the climate changers. This market is something like the grain or stock markets.

As recently as September 2022, an article in the Wall Street Journal encouraged the World Bank to put up millions of dollars for climate change activities (CO_2 controls).

- CO_2 in Air / Photosynthesis -

CO_2 is the feed stock for photosynthesis. This is the process where sun energy (sunlight) produces all plant life, which feeds humans and other living beings. These crops like wheat, corn, oats, soybeans, and others are the fundamental food for humans and animals. As a former farmer, I know they are very necessary for human life.

The earth is a gigantic energy engine, primarily run on sun energy (light). CO_2 is in the form of limestone (calcium carbonate). Limestone is generously distributed around the earth. It slowly dissolves in the sea water to give carbonic acid ($H_2O + CO_2 \rightarrow H_2CO_3$).

- CO_2 in Humans -

In humans, like in other animals, CO_2 is the result of food being processed in their bodies for the energy needed to sustain their life activities. A significant function of CO_2 is to maintain the blood's acidity level (pH) within a tight, critical range needed for all the body chemistry going on. If the blood acidity is off by even a little bit, the brain is altered or death happens. I had a niece that this happened to, and she became a "basket case," living but with no obvious physical ability or brain function.

- CO_2 and World Function -

I observe that the world is a giant energy machine. It has volcanoes, plate tectonics, vast storm systems, continuous electric discharges, lightning, and so on. It has many interacting processes going on that feed back to maintain and control the earth and its atmosphere within limits for life for many years. Everything is in constant change. This includes the sun. Humans cannot fight nature – trying to deliberately change the climate by controlling CO_2 is fruitless.

- Zero CO_2 Discharge -

If climate change people want zero discharge of industrial carbon dioxide (CO_2), they should push for the use of thorium (Th) energy. Compared to current atomic electric generation with uranium/plutonium processes, thorium energy is very simple to use, very safe, very plentiful, and lends itself to small operating units. For example, this could provide small thorium-powered factories all over the country, making the electric grid obsolete. The grid is very vulnerable to local problems and particularly national disasters such as hurricanes or accidents.

- Warmer Earth Temperature in the Past -

Historically, in 1421 (see the book *1421: The Year China Discovered America* by Gavin Menzies), the earth was much warmer than now. The book describes how the Chinese, by exploring the world to define longitude, found they could go much closer to Antarctica! They also sailed around Greenland, mapping about 80% of its shoreline. In that book, see the maps they made – their map of Greenland almost perfectly matches present satellite pictures of Greenland from space.

As far as climate is concerned, the world has a long way to go to be as warm as in 1421.

- Supercritical CO_2 (Carbon Dioxide) – Fluid CO_2 -

CO_2 in the supercritical state acts as a fluid. To achieve the supercritical state, CO_2 is under about 40 atmospheres of pressure. It is used extensively today in fracking for producing oil and gas from the earth. Fluid CO_2 is also used in various technologies, such as for analysis and for separation of chemicals.

Analytical Application: I have used fluid CO_2 for analysis of the monomer associated with a lubricant used for 6,6-nylon yarn spinning. First, we tried conventional analysis technology and got poor sensitivity. To get reasonable sensitivity, we sampled air around the nylon spinning for about 48 hours. This was very complex, since a sampling went through two 8-hour work shifts. We could report (as I remember) about less than 0.5 parts per million (ppm) of the candidate monomer pollutant.

At that time, the company (DuPont) owned a petroleum company, so I got their analytical section to help develop a substantially more sensitive method using fluid CO_2 and chromatography technology. With the new method, we could report monomer air contamination in the parts per billion (ppb), using only about an hour of sampling.

Stereoisomer Applications: Fluid CO_2 is used to separate stereoisomers of organic compounds. This is a very sensitive method, since stereoisomers are chemicals that are the same except that they are mirror images (like your right hand is the mirror image of your left hand). Why is this important? Because almost all life chemistry is the levo (left-handed) not the dextro (right-handed) form. This means that the levo mirror image isomer causes plain polarized light to rotate to the left, while dextro rotates polarized light to the right.

Although human life is made up of levo stereoisomers, chemicals and pharmaceuticals prescribed for them are racemic mixtures, equal parts of levo and dextro stereoisomers. It may be that the levo and dextro versions of a pharmaceutical can have effects that are the same, similar, opposite, or completely different. Both may be helpful in the same or different ways, or one may be toxic.

There is a lot of chemistry going on to develop chemical processes to give only levo-isomers or only dextro-isomers as pharmaceuticals. Of course, at least now we know how to separate the isomers, by fluid CO_2 chromatography.

Other Fluid CO_2 Uses: Fluid CO_2 is a general solvent for organic compounds. This is an alternative for toxic solvents to do chemistry in, for example, benzine. Of course, the price to pay is working at about 40 atmospheres of pressure.

- References -

For worldwide extensive studies of carbon dioxide in air and oceans, see this report: By Sofie Bates, NASA's Earth Science News Team. (2021, December 8). NASA-supported study confirms importance of Southern Ocean in absorbing carbon dioxide. Climate Change: Vital Signs of the Planet. https://climate.nasa.gov/news/3136/nasa-supported-study-confirms-importance-of-southern-ocean-in-absorbing-carbon-dioxide/

RISK/SAFETY

While I worked in industry, I was on a process hazards committee. We investigated any change in our company's processes, proposed or actual, to judge its safety issue. Here are a few examples of what this committee dealt with.

- Heat-transfer Process Rolls -

An example of the committee's role was with heat-transfer process rolls. They were made in Germany and shipped by air to the US. The process rolls were of steel, with a cavity inside, filled partly with water as a heat-transfer liquid.

The most important question was: Were they shipped in the storage space of the plane, which was not temperature- and pressure-controlled during flight?

They were, so we rejected them as being unsafe. The reasoning was that the water froze during the flight, and probably water expansion cracked the metal slightly. Note: Water freezing and thawing in/on the hardest rock, even granite, eventually turns it to small particles. (The question I had: Why did they use water (a lousy heat-transfer medium) rather than better heat-transfer fluids such as "Dowtherm A"?)

- Polyurethane -

My company's engineers tried to develop melt spinning of polyurethane; when they came to the process hazards committee for safety approval, we of course said no. We explained to them that if polyurethane is heated, it gives off a poisonous gas. After inquiring if anyone had died in their early tests, the answer was no. Apparently, the research lab was well ventilated.

- Fluorochemicals / Anti-soil Yarn -

Another request for safety clearance was to approve a fluorochemical. Since this ingredient was made by my company, I found out how it was made and its composition. At first, I declared it unsafe because the ingredient had maybe one or two percent of a very flammable solvent that was used to put the fluorochemical into a micro-dispersion. I calculated there was enough of the solvent so that as it ran, it would be evaporated. Its vapors would first burst into flames, then explode, sparked by the open electric motor in the process. To correct this problem after the ingredient was made, they could distill off the solvent. This was a very simple thing to do, so they did it.

Later, at an interdepartmental review, I described the whole adventure. After a little bit, I asked the people that made it if they corrected all the ingredient, including the ingredient sold to other companies. The answer was no. Then I asked if that ingredient had burned down customer factories. There was complete quiet for a bit; obviously they were seeing the significance of the unsafety of the ingredient.

MYTHS

- Indian Lore -

I read a lot of history and other information about American Indians (Native Americans) – for example, history books by Allan W. Eckert. He announces in each book that he only writes that which he can confirm by Native Americans and by US Army records. For the Battle of Fort Dearborn (today, the Chicago area), most reports in newspapers and historic books are based on descriptions by an American in the middle of the battle. His comments are discounted because of the fact that during the battle, he hid in an outhouse with a half-moon door window. Therefore, his reports are not believed by historians, since he had a very limited view and could not have seen all that was attributed to him.

- Mandan Indians -

I was born in North Dakota in the area of the Mandan Indians. They were noted for supplying Sacajawea as language interpreter for the Lewis and Clark expedition to the Pacific Ocean. The Mandan myth is about their honoring new babies that arrived with white skin, blue eyes, and red hair. In addition, they honored the Blessed Virgin Mary, the mother of Jesus. I read a book that was to explain all that: Irish and Scottish clergy and lay adventurers travelled to North America well before Columbus, travelling down the East Atlantic coast, through the Caribbean to the Mississippi River, and up the Missouri River to what is now North Dakota and the Mandan Indian lands.

They married the locals and distributed their characteristics among the Mandan. Much later, the French came there by way of the St. Lawrence River, then through the Great Lakes to the Midwest. Still later, various Europeans from the Eastern US arrived. These people reported the myths discussed above.

- Weird Ships and People – Indian Lore -

In my reading about American Indians, there were a number of reports along the East Coast of North America about great sailing ships and strange people. The people wore a single piece of clothing that covered them from shoulders to ankles.

I learned later that it probably was the Chinese. In the book *1421: The Year China Discovered America* by Gavin Menzies, it described their sailing northward along the North American coast. Their ships were as big as current ships, particularly ones that carry petroleum around the world.

- Vikings in Minnesota -

Growing up in Minnesota, I often heard that the Vikings visited there in about 1000 AD. The evidence was a "rune" stone with Viking markings on it, and other marker stones around the state. The story was that after getting to Vinland, they continued west through the St. Lawrence River and the Great Lakes into the country beyond Lake Superior. The story was perpetuated by the Norwegians who settled in the state.

- The Sunken Ship -

In the early 1900s (as I remember), a farmer on the Eastern Shore of Maryland decided that he would build a small irrigation pond to irrigate his truck garden produce, such as melons, strawberries, potatoes, and others. The task was to dig a hole about 20 or 30 feet deep. Generally, the water table is near the surface, so soon the hole would be naturally filled with water to make a small pond.

When he got the hole dug, and before it filled with water, he had come across a sailing ship embedded there. When I heard that, I asked if he would mind if I got a friend who was a scuba diver to identify the ship (an archeological find). No, he said, wait for a dry spell and the pond goes dry. The archeologist could just go at the dry hole with his hammer and trowel, and shovel at his leisure.

- Iraq War -

There was a story going around in the second Iraq War that a tank was disabled by a very small straight hole put in it. The hole went through its armor on both sides and the engine. (No one was killed.) The thought was that it was done by a new weapon, maybe a very powerful laser.

- Methane Hydrate -

When plants and animals die, and they accumulate on the bottom of lakes, seas, and oceans. They degrade there under great pressure and no oxygen, and they form methane, a gas. If the temperatures and pressures are right, methane then forms methane hydrate by incorporating with water. This is a liquid, and being denser than water, it accumulates on the bottom of lakes, seas, and oceans. If the temperatures or pressures rise, methane hydrate separates into gaseous methane and bubbles and rises up to the water surface.

The myths of abrupt unexplained loss of planes and ships off the coast of Florida in World War II may be explained by conditions of methane bubbling to the water surface. Imagine, suddenly a plane or ship trying to continue to travel in the air or water among gross bubbles of methane and seawater; they are abruptly lost.

Another myth is where a deep African Lake surrounded by cliffs erupted, killing all the people on its shore. This was attributed to methane hydrate breaking up into gaseous methane foam and water, excluding oxygen (in air) and killing them. A report years later claimed it was carbon dioxide (CO_2). This is very questionable. How could that be CO_2?

By the way, here's another technology to-do: There are deep pools of methane hydrate in the deep oceans. We should secure this and use it as we do petroleum. It should be easier than getting petroleum via wells.

- Ship Looking for Manganese Nodules on Seabed -

There was a myth or government lie that the ship was not searching for manganese but was reading messages on copper cables between continents, reading electric emission on the surface of cables to see what other nations were discussing – in secret, supposedly.

NORTH AMERICA TAKEN OVER

North America was populated with millions of people before it was invaded by Europeans. It is interesting that they had developed along similar lines as in Europe and Egypt in Africa. Corn was their staple food, but also potatoes, tomatoes, melons, squash, and pumpkins. There were many wild animals for protein. They had developed great civilizations, as shown by earthen mounds. An example is a mound six miles east of the St. Louis Arch, near the junction of the Mississippi and Missouri Rivers.

In the period after Columbus, Europeans started to come to North America and brought with them smallpox. Europeans had quite good immunity to that, but Native Americans essentially had none. Therefore, probably 80 to 90% of Native Americans died off and their civilizations were decimated.

The natives had kept the wild animals controlled by hunting them for meat (protein) for their diet. With most of the natives gone, after a few decades without millions of people hunting them, the animal numbers grew rapidly. People from Europe arriving later reported so many buffalo that if one stood on a bluff, one could see the great herd moving to new pastures, going by for hours. Again, this is what Europeans saw in North America, great herds of animals, and hardly any native people about the country, much land that had been farmed for eons, and now was mostly empty for the taking. And they did.

Of course, there were still enough natives left behind that they fought back against the European invaders, but they were overrun by the vast influx of "foreigners."

A STORY FROM WORK

At my company, I often would have lunch in the cafeteria with young PhD lady hires. I asked if anyone had a near-death experience. When someone would tell me of their own near-death experience, I would respond, "You want to bet?"

I would then ask how life had been since then. If they said life had been great, my response was, "Maybe you actually died and you went to heaven." If their reply was that everything always went wrong, my response was, "You are getting a taste of hell." They did not like to hear that. My idea was to encourage them to think about life and their relationship with God.

God is an infinite God. He is infinitely loving (of his creations), infinitely just, and infinitely merciful. He does not condemn us to heaven or hell. We are made in his image and likeness with free will, so we choose at the many crossroads in life to go to heaven or to hell.

Even some of the worst sinners we know about, like St. Paul or Mary Magdalene mended their ways. St. Paul was struck blind, and Jesus asked him, "Why did you persecute me?" With Mary Magdalene, she was caught in prostitution. Jesus asked her accusers, "You without sin cast the first stone," and one by one they left, and Mary was told by Jesus to go and sin no more.

CHRISTIAN ETHICS – A REVIEW

Some problems can be solved by technology or political will. But to solve any problem, I recommend you understand what God wants you to do first.

Pride is probably the worst of the capital sins. If you consider yourself better than others, or you are of an elite group, I think that is pride. When I worked for DuPont, its philosophy was that all people have about the same brain capability; it just has to be developed. That is up to the individual, with the help of his/her associates. People are social, as God created them. In the following, I have tried to show the do-nots in a natural and Christian society.

But first, let me describe things to read, so you will understand how Christian philosophy developed from natural laws, as shown by Plato and Socrates, St. Thomas Aquinas and St. Augustine.

1. If one reads Plato and Socrates, they told of natural law.
2. Later, St. Thomas Aquinas explained that natural law was adapted into Christian law.
3. Also, read St. Augustine, who further explained this stuff.

- The Things to Avoid in Christian Philosophy -

First: God created humans in his own image and likeness. That makes it tough for us humans. We have free will.

We are being told by the Evil One that bad stuff is good. In my opinion, we are inspired by the Holy Spirit (God) to do good things. We have to choose. In so doing, we choose Heaven or Hell for ourselves. God does not condemn us nor give us a free ticket to Heaven.

Just think of this: Jesus said I am in you, and you are in Me, just as I am in the Father (God) and He is in Me.

So we are Jesus's hands and feet to do God the Father's will.

It is kind of like the family, as an example. Children are taught to do what their father says.

- The Do-Nots -

To begin this section, what is a mortal sin?

- It has to be something very bad.
- One has to do it with full knowledge.
- One has to have full consent of one's will.

(With these requirements, I wonder how many mortal sins there are, since many do things without thinking!)

- Capital Sins -

The capital sins are called that, not necessarily because they are worse than other sins, but because they are the bases of other sin. A capital sin may be a mortal sin, but not necessarily.

(This is from Catechism "101".)

- Pride

- Covetousness
- Lust
- Anger or rage
- Greed
- Gluttony
- Sloth (laziness)

- Ten Commandments -

1. I am the Lord, your God. Do not have false gods.
2. Do not take God's name in vain (swearing).
3. Keep holy the Sabbath.
4. Honor your father and mother (and parents, honor your children).
5. Do not murder.
6. Do not commit adultery.
7. Do not steal.
8. Do not bear false witness.
9. Do not covet your neighbor's wife.
10. Do not covet your neighbor's goods.

- The Problems -

- People abruptly act without thinking of the consequences. People should act with due diligence.
- To err or to sin often gets to be a habit. Correct or change from bad habits.

- God created us with strong instincts; these are good. But do not live by instincts. Where does instinct end and good morality begin? We are humans, we are not just animals. (Animals mainly live by instincts.)
- Humans often choose poorly, such as borrowing money, and then they are forced by the lender into un-Christian stuff. Choose a lender wisely. Avoid being coerced or forced to do wrong. One does not have to do wrong, forced by others. Stand up for your Christian ethics.
- Beware of false preachers. For example: Preaching "Thou shalt not kill." Historically, for many years, the commandment was "Thou shalt not murder." If one believes in "Thou shalt not kill," then guns would be outlawed, since their function is to kill. We need police and armies to protect us.

DOG STORIES

- Stray Dog -

When we lived in Bellingham, Minnesota, I slept on the open porch upstairs, and there was a great commotion outside at dawn. Apparently, a stray dog had come into our yard and was chasing chickens and making a large noise, barking and so forth. By the time I got out of bed, I saw my dad coming out of the house with his shotgun, and he decided not to shoot the dog but to shoot down into the gravel by the house, which kind of bounced and skinned the dog on one side. The dog abruptly ran off, howling continuously and racing rapidly out of sight and earshot. My dad told me afterwards that he was concerned that if he shot the dog, and if there was an owner, he would come and give him a hard time about killing his favorite dog. So he said that if the dog came home bruised and the owner complained about it to Dad, he would charge the owner $15 for each of the chickens that were killed.

- Bad Dog -

Our four-month-old puppy one summer afternoon went after my mother's chickens that had gotten loose from the chicken pen, and killed one after another. With all the noise, my mother came out, took a broom and chased and caught the dog, and tied it up by the barn. When Dad came home, he was directed by Mom to take care of the dog. Dad got a gunny sack, put two dead chickens in it, put the dog in it so his nose was in amongst the dead chickens, then whipped him with a folded newspaper until he was yelping and howling, then let him go. Thereafter, the dog ran whenever a chicken came close to him. My dad explained: He first thought to get the shotgun and blow the dog to smithereens, but then he thought the dog just did what came naturally to him. This way the dog learned a lesson he never forgot, and could continue to be our family dog.

- Grandpa's Dog? -

Grandpa Furia was a dog lover. When he died, we had a funeral and buried him. He didn't have a dog in later life. We stayed over the weekend. Barbara said we should go to the cemetery and say a prayer. But when we arrived, we saw there was a dog on the pile of dirt that was over Grandpa Furia's grave. He was protecting it and wouldn't let us near.

- Smart Dog -

Next door to us in Salisbury was the Guy family. Their youngest son, David, was about the same age as our daughter Carol. Their yard didn't have a fence around it. But the dog knew that David wasn't allowed outside the yard. So when the little boy, who was maybe two years old, would start out of the yard, the dog would take him by his shirt and put him back in the grass.

In later years, I was talking to David's mother. She said the dog had another thing that he did. When the children were playing in my yard, we had a fence around the yard but the gate was rotted away so that the kids weren't really confined to the yard. But the dog would stand at the gate and growl at the kids, assuring they would not get out of the fence.

- Dog Language -

In the Grandpa Peter Hackert household, there were nine kids, five by Grandma Martha, and four by Grandma Ida. The school teacher for the country school across the road also lived with them. The talk around the supper table involved three languages—classical German, Low German, and English. If anyone was talking to the teacher, it was in English. When Grandpa talked to his sons, it was in Low German. Grandma Ida only knew classical German, so when she talked to the daughters or others, she spoke classical German.

One winter day, Grandpa Peter was going to Rosen with four-year-old Wilbert in their horse-drawn wagon. On the way, they passed by a farmplace, and its family dog came out and pestered the horses, so Grandpa Peter yelled at the dog. After a little bit, the dog went back home. Little Wilbert asked his pa, "How did you know that was a boy dog?" Obviously, his pa must have been yelling at the dog in Low German, the language of boys!

www.ingramcontent.com/pod-product-compliance
Lightning Source LLC
LaVergne TN
LVHW010630100826
845148LV00014B/3183

* 9 7 8 1 6 2 8 0 6 4 7 0 4 *